PLANET MARKU

 A discovery activity
by Samuel Marcus and
Robert (Boppa) Harris

<u>Your Mission, Should You Choose to Accept It</u>

You have the opportunity to create your own planet! This is an interactive discovery activity that will result in your own special planet. In fact, this book provides you with three blank (but guided) pages for each section of your book in which to describe your planet. You will have the opportunity to create three unique planets. We have provided you some guidelines, but you can research the kinds of things other planets have and add these special features to your planet.

<u>The Name of the Planet</u>

The name "Planet Marku" comes from a fun pronunciation of the author's last name, Marcus. Choose a fun name for your planet --- a relative's name, a friend's name, your pet, or just something with which you want to have fun.

<u>The Moons of Planet Marku[1]</u>

- Yankee
- Unicorn
- Derek
- Sam
- Boppa
- Eriksen
- Spurs
- Judge
- D@ BEEEST

<u>Time on Planet Marku</u>

- A day on Planet Marku is 17 hours
- A year on Planet Marku is 779 days

<u>The Rings of Planet Marku</u>

- The rings that surround Planet Marku are neon green and neon yellow from the reflection of sunlight off space junk. They are made of green and yellow space "junk" consisting of green and yellow marker caps.

---

[1] Note the personal nature of these moons --- Sam is a big soccer and New York Yankees fan --- readers will be encouraged to use names of special meaning to them.

<u>The Atmosphere of Planet Marku</u>

- Planet Marku has the same temperature range as Earth and is the second known planet in the universe to support life.
- Planet Marku has 3/17 the gravity of earth.
- There are three layers:
    - The Wondosphere: Hot and cold air connects (Storms)
    - The Smellosphere: Full of gases
    - The Weathosphere: Where the clouds are (No Storms)

<u>The Surface of Planet Marku</u>
- Planet Marku has a great neon pink spot and a crater as big as Jupiter. Planet Marku is made of the same material as a bouncy ball and is covered with stars which make it very bright, like our sun. There is one big orange ocean and six continents.
- There are many (blue/purple) plasma eruptions (same color as bouncy ball material) on the surface of Planet Marku.

<u>The Exploration of Planet Marku</u>

- The Space Probe Pioneer 10 is very close to Planet Marku. We have figured out that Pioneer 10 passed by Planet Marku on October 21, 2015.
- The Voyager Space Probe has also come very near to planet Marku (50,000 miles away, on January 30, 2016).

NAME OF MY PLANET_____

COVER GRAPHICS

AUTHORS

HOW I WANT MY COVER TO LOOK

NAME OF MY PLANET_____

COVER GRAPHICS

AUTHORS

HOW I WANT MY COVER TO LOOK

NAME OF MY PLANET_____

COVER GRAPHICS

AUTHORS

HOW I WANT MY COVER TO LOOK

The Moons of Planet Marku

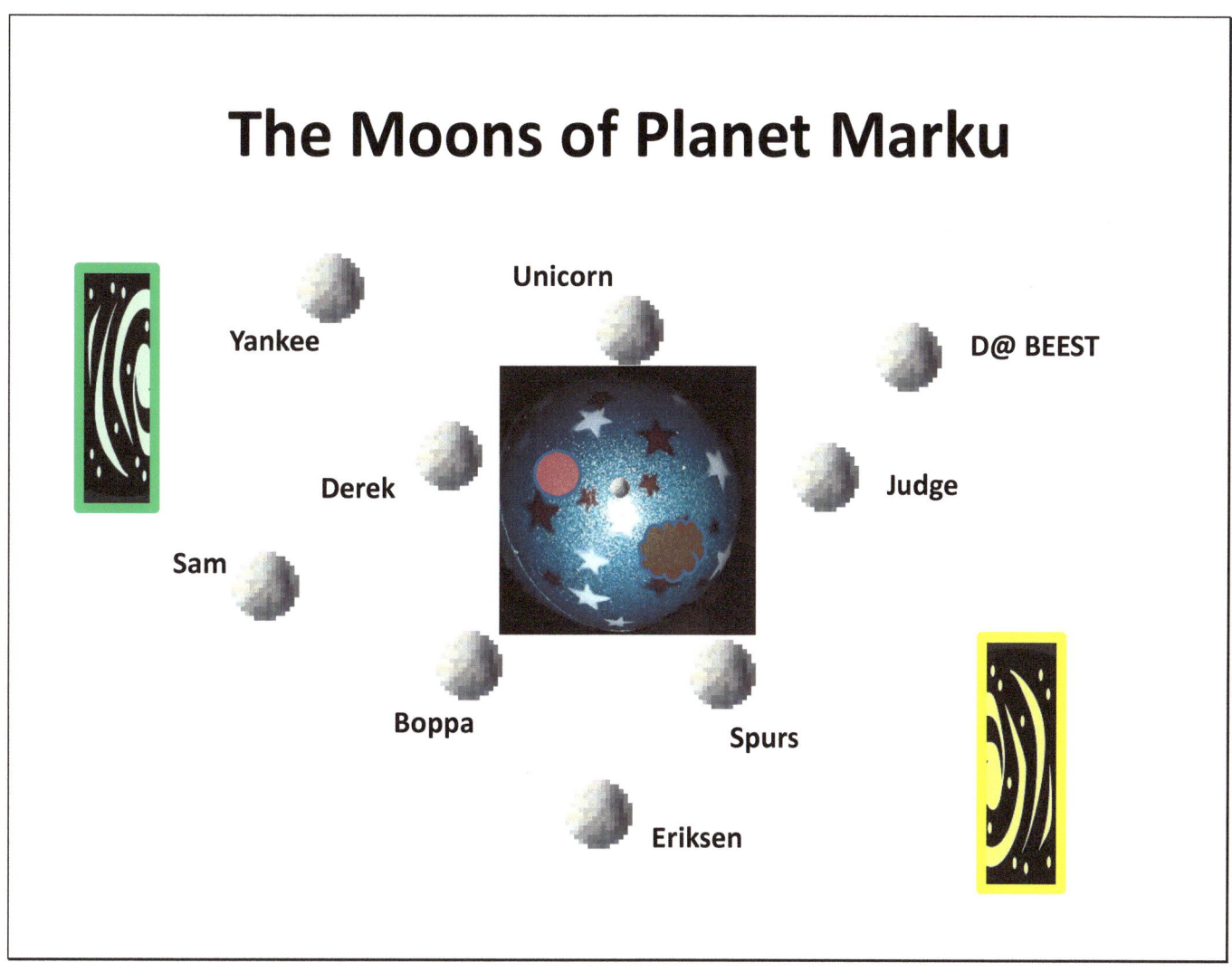

NAME OF MY PLANET_____

NUMBER OF MOONS _____

NAMES OF MOONS

HOW I WANT MY MOON PAGE TO LOOK

NAME OF MY PLANET_____

NUMBER OF MOONS _____

NAMES OF MOONS

HOW I WANT MY MOON PAGE TO LOOK

NAME OF MY PLANET_____

NUMBER OF MOONS _____

NAMES OF MOONS

HOW I WANT MY MOON PAGE TO LOOK

Time on Planet Marku

A day on Planet Marku is 17 hours

A year on Planet Marku is 779 days

NAME OF MY PLANET_____

TIME ON MY PLANET

- **HOW LONG IS A DAY?**
- **HOW LONG IS A YEAR?**

HOW I WANT MY TIME PAGE TO LOOK

NAME OF MY PLANET_____

TIME ON MY PLANET

- ## HOW LONG IS A DAY?
- ## HOW LONG IS A YEAR?

HOW I WANT MY TIME PAGE TO LOOK

NAME OF MY PLANET_____

TIME ON MY PLANET

- ## HOW LONG IS A DAY?
- ## HOW LONG IS A YEAR?

HOW I WANT MY TIME PAGE TO LOOK

The Rings of Planet Marku

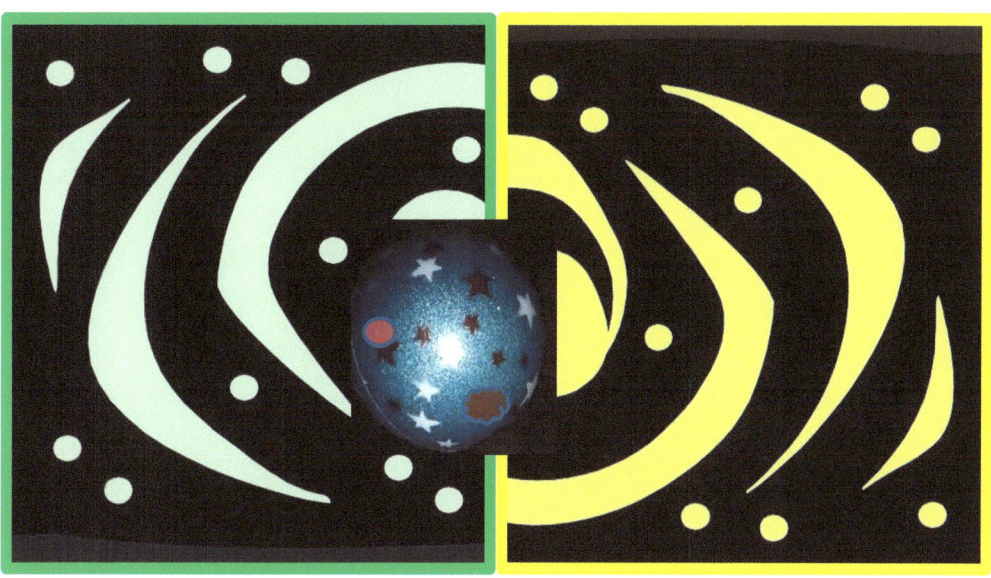

The rings that surround Planet Marku are neon green and neon yellow because of the variation in how sunlight reflects off the space junk. They are made of green and yellow space "junk" consisting of green and yellow marker caps.

NAME OF MY PLANET_____

THE RINGS OF MY PLANET

ANY OTHER ORBITING BODIES BESIDES MOONS?

HOW I WANT MY RINGS PAGE TO LOOK

NAME OF MY PLANET_____

THE RINGS OF MY PLANET

ANY OTHER ORBITING BODIES BESIDES MOONS?

HOW I WANT MY RINGS PAGE TO LOOK

NAME OF MY PLANET_____

THE RINGS OF MY PLANET

ANY OTHER ORBITING BODIES BESIDES MOONS?

HOW I WANT MY RINGS PAGE TO LOOK

The Atmosphere of Planet Marku

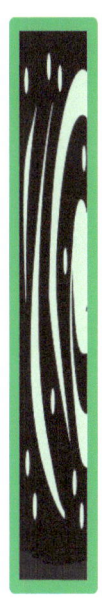  

Planet Marku has the same temperature range as Earth, and is the second known planet in the universe to support life. Planet Marku has 3/17 the gravity of Earth.

There are three layers: Wondosphere (Hot and cold air – Storms); Smellosphere (Full of gases); and, Weathosphere (clouds – No Storms)

NAME OF MY PLANET_____

SPECIAL ATMOSPHERIC FEATURES OF MY PLANET
- **TEMPERATURE?**
- **GRAVITY?**
- **LAYERS?**

HOW I WANT MY ATMOSPHERIC FEATURES PAGE TO LOOK

NAME OF MY PLANET_____

SPECIAL ATMOSPHERIC FEATURES OF MY PLANET

- TEMPERATURE?
- GRAVITY?
- LAYERS?

HOW I WANT MY ATMOSPHERIC FEATURES PAGE TO LOOK

NAME OF MY PLANET_____

SPECIAL ATMOSPHERIC FEATURES OF MY PLANET
- **TEMPERATURE?**
- **GRAVITY?**
- **LAYERS?**

HOW I WANT MY ATMOSPHERIC FEATURES PAGE TO LOOK

The Surface of Planet Marku

Planet Marku has a great neon pink spot and a crater as big as Jupiter. Planet Marku is made of the same material as a bouncy ball and is covered with stars which makes it very bright, like our sun. There is one big orange ocean and six continents. There are many (blue/purple) plasma eruptions (same color as bouncy ball material) on the surface of Planet Marku.

NAME OF MY PLANET_____

SURFACE FEATURES OF MY PLANET
- SPOTS?
- CRATERS?
- MATERIALS?
- OCEANS?
- LAND MASSES?

HOW I WANT MY SURFACE FEATURES PAGE TO LOOK

NAME OF MY PLANET_____

SURFACE FEATURES OF MY PLANET

- SPOTS?
- CRATERS?
- MATERIALS?
- OCEANS?
- LAND MASSES?

HOW I WANT MY SURFACE FEATURES PAGE TO LOOK

NAME OF MY PLANET_____

SURFACE FEATURES OF MY PLANET

- SPOTS?
- CRATERS?
- MATERIALS?
- OCEANS?
- LAND MASSES?

HOW I WANT MY SURFACE FEATURES PAGE TO LOOK

The Exploration of Planet Marku

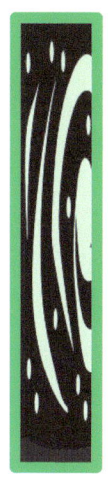  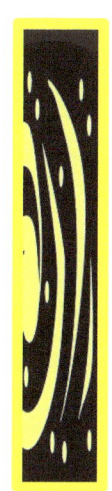

The Space Probe Pioneer 10 is very close to Planet Marku. We have figured out that Pioneer 10 passed by Planet Marku on October 21, 2015. The Voyager Space Probe has also come very near to Planet Marku (50,000 miles away, on January 30, 2016.)

NAME OF MY PLANET_____

EXPLORATION OF MY PLANET

- HAVE ANY COUNTRIES ON EARTH EXPLORED MY PLANET?
- WHAT HAS BEEN DISCOVERED SO FAR?
- IS THERE LIFE ON MY PLANET?
- WHEN IS THE NEXT PLANNED SPACE LAUNCH TO MY PLANET?

HOW I WANT MY EXPLORATIONS PAGE TO LOOK

NAME OF MY PLANET_____

EXPLORATION OF MY PLANET

- HAVE ANY COUNTRIES ON EARTH EXPLORED MY PLANET?
- WHAT HAS BEEN DISCOVERED SO FAR?
- IS THERE LIFE ON MY PLANET?
- WHEN IS THE NEXT PLANNED SPACE LAUNCH TO MY PLANET?

HOW I WANT MY EXPLORATIONS PAGE TO LOOK

NAME OF MY PLANET_____

EXPLORATION OF MY PLANET
- HAVE ANY COUNTRIES ON EARTH EXPLORED MY PLANET?
- WHAT HAS BEEN DISCOVERED SO FAR?
- IS THERE LIFE ON MY PLANET?
- WHEN IS THE NEXT PLANNED SPACE LAUNCH TO MY PLANET?

HOW I WANT MY EXPLORATIONS PAGE TO LOOK

WE HOPE YOU HAVE HAD

FUN WITH YOUR PLANETS !

PLEASE FEEL FREE TO

CONTACT THE AUTHORS OF

PLANET MARKU

AT

RHARRISPHD@GMAIL.COM

www.ingramcontent.com/pod-product-compliance
Lightning Source LLC
Chambersburg PA
CBHW050421180526
45159CB00005B/2356